#STANDARDS tweet

140 Bite-Sized Ideas on Winning in Technical Standards

By Karen Bartleson

Published by THINKaha™, a Happy About® imprint 20660 Stevens Creek Blvd., Suite 210, Cupertino, CA 95014
http://thinkaha.com

First Printing: December 2010
Paperback ISBN: 978-1-61699-014-5 (1-61699-014-7)
eBook ISBN: 978-1-61699-015-2 (1-61699-015-5)
Place of Publication: Silicon Valley, California, USA
Paperback Library of Congress Number: 2010938215

Advance Praise

"A career's worth of standards wisdom in 140 characters or less."
@ronploof, Author of *Read This First: The Executive's Guide to New Media*

"This book emphasizes the usefulness of standards and their critical role in today's day and age."
@arrafiq, Consultant at Inspurate Design

"In STANDARDStweet @KarenBartleson gives you all you need to know to be a success in the all important Standards Game. Read it and prosper."
@richgoldman, Vice President of Corporate Marketing and Strategic Alliances at Synopsys, Inc.

"This book conveys in 140 tweets of 140 characters or less pearls of wisdom it's taken me over two years of committee meetings to learn!"
@jlgray, Vice President at Verilab

"Insightful advice for all standards developers from the high priestess who has preached this for 20+ years. Thanks for sharing."
@ytrivedi, Standards Director at Synopsys, Inc.

Dedication

To everyone who has played, plays, or will play the standards game.

Acknowledgments

Many thanks to all my colleagues in the world of standards. You've taught me so much and given me countless opportunities to grow.

A special thanks to Jack Dorsey, Biz Stone, and Evan Williams for creating Twitter which brought an amazing communication channel to people everywhere. I'm Twitterpated.

Finally, I appreciate the support and encouragement from the folks at THINKaha and Happy About who motivated me to create *#STANDARDS tweet*. You all are great!

Why Did I Write This Book?

Standards have played a large part in my career.

I enjoyed writing the book, *The Ten Commandments for Effective Standards.*

"STANDARDStweet" is a nice companion to *The Ten Commandments for Effective Standards.*

"STANDARDStweet" can be read in very little time.

It's satisfying to share my "lessons learned" with others.

The standards game is not for sissies.

Standards are important to everyone, everywhere, whether they realize it or not.

Snippets of standards experience might make people's standardization jobs easier.

Twitter is my favorite social network.

I have lots of standards friends on Twitter.

Combining standards and Twitter was a kick.

Being a book author is a rewarding opportunity of a lifetime.

Karen Bartleson

Contents

Foreword

Technical standards affect us every day. They are the secret ingredients that make electronic products work together. Engineers spend their entire careers working with standards, but as I've discovered, working with standards and creating them are two entirely different things. In "STANDARDStweet," standards guru Karen Bartleson captures a wide range of thoughts that help put technical standards into proper perspective.

JL Gray is Vice President of Verilab, a consulting firm that provides services for advanced computer chip design. As another pioneer in social media for his industry, he's the author the popular blog, "Cool Verification." He's a standards contributor and active on Twitter. His Twitter name is @jlgray.

See how easy... plug 'n play!
Nothing like a solid standard that really works

Section I

Why Standards?

Standards are part of life. If you look, you'll see them every day all around you. Technical standards make things work together. They are good for businesses and consumers alike.

1

Standards are all around us. They serve a role in just about everything we use, eat, sell, and enjoy.

2

Simply put, a standard is everyone doing something the same way.

3

When things work seamlessly, typically there are standards at play. It's just that they become "invisible" when implemented right.

4

Think standards if you want to improve quality, lower costs, and streamline the development cycle of products.

5

Think standards if you want to help protect our safety, health, and environment.

6

A standard's value is directly linked to how much it helps a market to grow.

7

Standards create open interfaces that vendors can use to make products work together better for customers.

8

Interface standards build networks and promote product sales. Not convinced? Imagine a world with a single fax machine...

9

Without standards, you might get a big share of a small market. With standards, you can get a share of a giant market.

10

"Plug and play" would be but a pipe dream if there were no standards.

11

Times are challenging. Interoperability – and suppliers' programs that promote it – are now as important to customers as ever.

12

The ONE reason to keep investing in interoperability programs: Your customers REQUIRE them.

13

You can fear your competitors and shy away from standards or you can love your customers and embrace standards.

14

You have options: you can work on establishing standards OR you can work on reinventing the wheel.

15

Implementing standards may not make you win in the long run but without standards, you're almost guaranteed to lose.

16

The fundamentals of technical standards development pertain to many industries.

Section II

The Ten Commandments for Effective Standards

Based on experience, here are ten "commandments" in 140 characters or less for effectively creating technical standards. No religious connotation is implied, of course.

What do you do?
Well, I'm on a standards committee...
I'm really into standards...
"Fly Me To the Moon" is one of my favorites!

17

Effective Standards Commandment #1: Cooperate on standards, compete on products.

18

Effective Standards Commandment #2: Use caution when mixing patents and standards.

19

Effective Standards Commandment #3: Know when to stop (and start).

20

Effective Standards Commandment #4: Be truly open.

21

Effective Standards Commandment #5: Realize there is no neutral party.

22

Effective Standards Commandment #6: Leverage existing organizations and proven processes.

23

Effective Standards Commandment #7: Think relevance.

24

Effective Standards Commandment #8: Recognize there's more than one way to create a standard.

25

Effective Standards Commandment #9: Start with contributions, not from scratch.

26

Effective Standards Commandment #10: Know that standards have technical and business aspects.

27

Read the book, *The Ten Commandments for Effective Standards*, available from Amazon and other outlets.

Section III

Cooperate, Compete, Starting, and Stopping

A few notes about the standards lifecycle.

28

Standards are about building a community of people willing to do something the same way while differentiating themselves.

Honey, are you tweeting again?
No, dear, I'm building community

29

Standards allow and even promote competition.

30

Step one in creating standards: Find other interested parties who are willing to put forth the effort to produce and adopt a standard.

31

Succeeding with standards requires a "contribution" mindset.

32

For standards to come about, people have to agree and cooperate, even when they are competitors.

33

Working with competitors to allow your customers to make their best choices brings a certain measure of health to your company.

34

A typical birth of a standard: a few people generate an idea, try it out, see its potential, and convince a broader group to develop it.

35

Collaboration and cooperation between competing companies and key customers can result in standards developed in record time.

36

Put aside competitive interests. Let intense competition for market share begin when a working standard is successfully completed.

37

A winning standard is a road towards an open, level playing field.

38

Standards battles add up to significant productivity hits and costly delays for everyone.

39

Learn to sell all stakeholders on the benefits of standards.

40

Almost always, there are two sides (or more) pushing for their way to become the standard.

41

Individuals in standards have personal agendas. In most cases though, the common agenda is what will benefit the individual members.

42

Internal squabbles, hidden agendas, and competitive interests have often delayed a standard's completion by months and even years.

43

As odd as it may sound, sometimes individuals or companies truly want a standard to fail.

44

Resist the urge to flame email reflectors; resolve issues locally within the committee instead.

45

A standards story is all about how the customers will win BIG because of the standards.

46

Everyone working on a standard has some kind of interest in it. No one would waste time on a standard that he/she didn't care about.

47

Standards can make a company a "vendor of choice" instead of a "vendor of no choice."

48

If suppliers and customers go on too many quests to address a problem, it results in confusion, frustration, and the demand for a standard.

49

You should start a standards effort soon but not too soon.

50

There's wisdom in knowing the right balance between standardizing too early and standardizing too late.

51

The time when creating a standard is not only appropriate, but essential, is when customers cry out for one.

52

The best way to create a standard is to start with donations, not from scratch.

53

Standards organizations typically solicit donations from industry, then combine and transform them into a mutually acceptable standard.

54

Employing a mature technology for standardization can reduce risk.

55

Standardizing on immature technology can pose the question, "How sure are we that this technology really works?"

56

The technology to be standardized shouldn't be obsolete. Standards are expensive and time-intensive to develop.

57

Respect the standards lifecycle. Know when and why to stop a standards effort if it's not progressing.

58

The key to success for standards is adoption: anyone can create a standard through any means, but if it's not used, it has little value.

Section IV

Openness

"Openness" is a big word with many implications in the standards game.

59

Standards range from completely closed to fully open.

60

Players in the standards game use the term "openness" to mean all kinds of things.

61

“Open” means everyone can participate without having undue restrictions imposed upon the participants.

Isn't it hard to express yourself in only 140 characters?
Not when your tweets are limitless
That's some calling plan!

62

Openness does not mean giving away your advantage to your competition.

63

Open and constantly improving standards tend to attract new users, new ideas, more products, and market growth.

64

Open standards can be managed by committees, companies, or even individuals. The key is that access to the standards is not restricted.

65

Closed, proprietary standards greatly reduce interoperability and competition.

66

A standard that is closed, overly controlled, or stagnant has diminished value and slows market expansion.

67

A “standard” open only to paying customers and select companies that are not direct competitors is a gated community—not truly open.

68

Control of standards ranges from control by a single individual to control by everyone.

69

Standards licensing schemes can be susceptible to confusing variations.

70

Closed proprietary standards, owned by a single company, are available only to that company's customers.

71

The closed proprietary approach has produced good business results, but this approach has had problems as well.

72

Open proprietary standards are controlled by a single company, and access is expanded to include the entire community.

73

Open proprietary ensures immediate access by everyone to well-established, well-maintained standards from the owner.

74

Community source licensing is a way of creating and evolving a standard while preserving the owning company's investment.

75

Open source standards are created, reviewed, and enhanced by a community of product developers and customers.

76

Open source standards are the fastest to create. They're managed by a single entity and everyone may suggest changes.

77

Formal standards can be slower to create, but they come with official accreditation that can lend more credibility to them.

Section V

Processes, Policies, and Procedures

Standards-setting is a team sport and when done right, there are multiple winners.

78

Regardless of what motivates a standards effort, it's important that a process is followed to ensure a successful outcome.

79

Standards-setting processes help ensure fairness—all participants have equal say in the resulting standards.

80

Technical expertise and business balance are essential to a successful outcome of a standardization effort.

81

The evolution of standards ranges from glacial to Internet speed.

82

One of the most important goals of producing a standard is to ensure that the standard comes out in a timely fashion.

83

The five values of the standards process: Due process, Openness, Consensus, Balance, Right of appeal.

84

Policies and procedures: Don't violate them!

85

Because standards are so important to a company's bottom line, standards bodies and processes are subject to close, public scrutiny.

86

Standards-setting processes may have holes that are discovered as the standards game ensues.

87

Following proven standardization processes is an efficient way to produce a standard.

88

Standards have to be maintained by a credible body that has experience in producing useful standards.

89

Standards bodies should continue to review and refine their policies and procedures as they learn from experience.

90

It can be tricky to maintain a balance of strictly-enforced rules and flexibility to accommodate modern business practices.

How many characters do I have left?
At this rate, not nearly enough
IEEE standard 1076/D4.2 (Revision of IEEE 1076-2002 specifies that...
A Standards Tweet

91

In setting standards, it's challenging to integrate different views and converge them into something that everyone will benefit from.

92

There is a cost for participating companies to adopt standards. If the costs are prohibitive, the companies will walk away.

93

The cost of committee participation may seem expensive, but if the result is a complete and fully adopted standard, it's worth it.

94

Standards can be created in a variety of ways, following different models.

95

There are many successful existing processes for standardization.

96

Some non-standard (pun intended) methods of standardization have been quite successful.

97

Look for an existing standards organization to create a new standard whenever possible.

98

When creating a new standards body, it's helpful to pattern policies and procedures after existing ones.

99

Formal standardization can be a lengthy process because all contingencies and considerations must be taken into account.

100

Formal standards organizations are good at managing stable, established standards that require a universal seal of approval.

101

Be realistic about schedules. New organizations almost always think they can create standards faster than has been done in the past.

102

A single standard is best, but it is also undeniable that two standards are better than three or ten or a hundred.

Section VI

Technology, Business, and Intellectual Property

Standards creation is multi-faceted with tension coming from different directions.

TOWER of BABEL REDUX

103

Real customer input is crucial to a good standard.

104

Proven technology forms the best foundation for a standard.

105

Take a practical approach to standards. Think about whether the standard will really be used, how it will be used, and by whom.

106

Adoption of standards ranges from narrowly used to pervasive.

107

There are some standards that have absorbed large amounts of effort, only to sit unused on the shelf.

108

An actual problem to be solved has to be identified to avoid a standard that goes unused or must wait for a problem to come along.

109

Useful standards are concise and do not include wish-list items from beyond their scope.

110

Standards have both technical and business aspects.

111

Don't underestimate the importance of the business aspects of a standardization effort.

112

Successful standards have benefits for all. Companies don't usually participate in activities that do not help their bottom line.

113

Before considering an open source standard for your product, it's a good idea to understand how it will be managed.

114

Standardize the interfaces between products, not the intellectual property inside them.

115

You can try to ignore legal issues surrounding standards, but legal issues surrounding standards will not ignore you.

116

Standards must not tread on a company's trade secrets, intellectual property, or proprietary technology.

117

Companies that own intellectual property of any kind are not likely to give it up simply for the "good of the industry."

118

Companies that built businesses on their technology shouldn't have to give it up in the name of standards if they don't want to.

119

If a company is unwilling to release its proprietary information, talk about it. Don't resort to smear campaigns or pressure tactics.

120

A patent pool can mitigate situations where companies desire to retain their patent rights while developing industry standards.

121

The concept of a patent pool is two or more companies agreeing to cross-license their patents that are related to a specific technology.

122

A patent pool is formed of "essential patents," which are those that would have to be infringed upon in order to implement the standard.

123

Standards are not the end. But they are a means to the customer's end benefit.

Section VII

Final Standards Tweets

Some additional thoughts on standards and their value.

124

Pay attention to standards. They can significantly impact - positively or negatively - our businesses and customers.

125

Standards move an industry forward.

126

Standards increase productivity and solve design problems while leaving plenty of room for innovation.

127

With the right standards, the marketplace expands, creating BIGGER opportunities for everyone who is participating.

128

Standards accelerate the possibility of innovation for the entire ecosystem.

129

Customers don't ask for standards. They EXPECT them.

130

Standards can create a positive and memorable user experience.

131

Proprietary technology can provide you a competitive advantage, but it might do so at the convenience of the customer.

132

With proprietary technology, you get to HAVE everything, but you get to do ALL the work to get it in the hands of your target market.

133

No customer wants to get the feeling that he or she is being held hostage by your proprietary technology.

134

We must produce our standards faster than our technology becomes obsolete—which is awfully fast in high tech.

135

Don't be surprised by the politics of the standards game.

136

All industries have "standards wars" because standards are so important to business.

137

You cannot NOT have standards. Life would be too difficult in a non-standard world.

138

Technology innovations will continue to present new opportunities for standardization.

139

Interoperability is a team sport and there are many winners in a standards game.

140

Have some fun while playing the standards game. You'll make good lifelong friends, even if they're your competition.

I've been retweeted,
therefore I am

About the Author

Karen Bartleson has three decades' experience in the computer chip industry. She is known for her work in the area of standards for electronic design automation and continues to serve on several committees that develop technical standards. She is the author of the book *The Ten Commandments for Effective Standards* and "The Standards Game," a blog focused on the standards arena. Karen holds a BSEE from California Polytechnic State University, San Luis Obispo, California. She was the recipient of the Marie R. Pistilli Women in Design Automation Achievement Award in 2003. Her Twitter name is @karenbartleson.

About the Cartoonist

Rick Jamison is a rare entity in the corporate world. By day, he's disguised as the mild-mannered Social Media Strategist at Synopsys (as well as Executive Editor of Synopsys Press Business Series). But at sundown, he reveals his real superpowers as an author and corporate cartoonist. Part illustrator, part subject clarifier, and part Big Business underbelly tickler, his words and cartoons enlighten, enliven, enrich, entertain...and from time to time even educate. His Twitter name is @rickjamison.

Other Books in the THINKaha Series

The THINKaha book series is for thinking adults who lack the time or desire to read long books, but want to improve themselves with knowledge of the most up-to-date subjects. THINKaha is a leader in timely, cutting edge books and mobile applications from relevant experts that provide valuable information in a fun, Twitter-brief format for a fast-paced world.

They are available online at http://thinkaha.com or at other online and physical bookstores.

1. *#BOOK TITLE tweet Book01:* 140 Bite-Sized Ideas for Compelling Article, Book, and Event Titles by Roger C. Parker
2. *#COACHING tweet Book01:* 140 Bite-Sized Insights On Making A Difference Through Executive Coaching by Sterling Lanier
3. *#CONTENT MARKETING tweet Book01:* 140 Bite-sized Ideas to Create and Market Compelling Content by Ambal Balakrishnan
4. *#DEATHtweet Book01:* A Well Lived Life through 140 Perspectives on Death and its Teachings by Timothy Tosta
5. *#DIVERSITYtweet Book01:* Embracing the Growing Diversity in Our World by Deepika Bajaj
6. *#DREAMtweet Book01:* Inspirational Nuggets of Wisdom from a Rock and Roll Guru to Help You Live Your Dreams by Joe Heuer
7. *#ENTRYLEVELtweet Book01:* Taking Your Career from Classroom to Cubicle by Heather R. Huhman
8. *#JOBSEARCHtweet Book01:* 140 job search nuggets for managing your career and landing your dream job by Barbara Safani
9. *#LEADERSHIPtweet Book01:* 140 bite-sized ideas to help you become the leader you were born to be by Kevin Eikenberry
10. *#LEAN STARTUP tweet Book01:* 140 Insights for Building a Lean Startup! By Seymour Duncker

11. *#MILLENNIALtweet Book01:* 140 Bite-sized Ideas for Managing the Millennials by Alexandra Levit
12. *#MOJOtweet:* 140 Bite-Sized Ideas on How to Get and Keep Your Mojo by Marshall Goldsmith
13. *#PARTNER tweet Book01:* 140 Bite-Sized Ideas for Succeeding in Your Partnerships by Chaitra Vedullapalli
14. *#PROJECT MANAGEMENT tweet Book01:* 140 Powerful Bite-Sized Insights on Managing Projects by Guy Ralfe and Himanshu Jhamb
15. *#OPEN TEXTBOOK tweet Book01:* Driving the Awareness and Adoption of Open Textbooks by Sharyn Fitzpatrick
16. *#QUALITYtweet Book01:* 140 Bite-Sized Ideas to Deliver Quality in Every Project by Tanmay Vora
17. *#SOCIALMEDIA NONPROFIT tweet Book01:* 140 Bite-Sized Ideas for Nonprofi t Social Media Engagement by Janet Fouts with Beth Kanter
18. *#SPORTS tweet Book01:* What I Learned from Coaches About Sports and Life by Ronnie Lott with Keith Potter
19. *#STANDARDStweet Book01:* 140 Bite-Sized Ideas for Winning the Industry Standards Game by Karen Bartleson
20. *#TEAMWORK tweet Book01:* 140 Powerful Bite-Sized Insights on Lessons for Leading Teams to Success by Caroline G. Nicholl
21. *#THINK tweet Book01:* Bite-sized lessons for a fast paced world by Rajesh Setty

www.ingramcontent.com/pod-product-compliance
Ingram Content Group UK Ltd.
Pitfield, Milton Keynes, MK11 3LW, UK
UKHW020134250726
13967UKWH00002B/654